Series 1

Lottery Pa[l]

The smart way to win Pick3, Pick 4 LOTTERY straight

> 657 straight on 7/10/09 DC
>
> 766 straight on 7/11/09 DC
>
> 170 straight on 7/7/09 DC
>
> 2165 straight on 2/17/09 DE
>
> 0333 straight on 8/24/09 DE
>
> All straight winnings on the LP

Eze Ugbor

I want to dedicate this book to my God for giving me wisdom & strength to make this book possible. I also want to thank my family, my wife Oge & my 3 children, Che-Che, Nana, & E.J. for being there with me patiently through this process. Finally, I would like to thank my mother for giving me guidance and teaching me how to stay focused even when things seem very difficult.

I thank you all!

Lottery Jargons

Words you may need to know

For the new subscribers that are not familiar with some of the words we use in the lottery world.

Lottery Pal: Series 1

<u>Straight/Exact</u> - this is when the number plays the exact position

Example: 234. If you play 234 and it comes out in exact 234 positions that is straight/exact position. In most states the payout for $1. Straight win is about $500.

<u>Box/Any other</u> - this is when the number comes out in any other. Example of this on 234 will be if it comes out in any other way than 234. The 234 has six different positions, 234, 342, 423, 432, 243 and 324. In most states the payout for $1. Box is about $80.

<u>Double number</u> - Example: 244. They can only come in three way positions: 244, 424 and 442. When you win the double

number straight you still get paid about $500 for every $1 wagered. However, when you win the double number box, you get double for every $1. That will be about $160.

LP - Lottery Pal

LP groups

Lead trigger/lead number   234

567

890

In the LP System the first number that plays when you are observing the group is the lead trigger or the lead number. That is the number you need to see before playing the other two numbers in that group. The LP number group is the three numbers that you are targeting. Any of the three numbers can become the lead trigger.

Triple star is when the number can only be won straight example 999.

LP System

From the LP series 1

339 is the lead indicator. This is the number that you have been waiting for to start playing. For those targeting to hit two which of course implies playing more, once the 339 drops, you take on the other two which will be 060 and 955. Sometimes the play will drop the third number before coming back to the middle number. The numbers are isolated to move in these groups. They move in tandem but the ultimate goal is to minimize the amount of money you spend playing and in most cases play with the house money and walk away a winner anytime you hit it big. The goal of the house is not to lose according to popular axiom and your goal should be not to lose. If you play with LP System, what stops you from aiming to play the role of the house? Who stops you from aiming to eliminate the word gamble? If you can play meticulously and come out a winner would playing not be fun.

You are now armed with the most potent system ever known to man. This is the first time this system is coming into the market. It will be series and there are only 10 series of the LP System if you want to cover every possible angle. Play with it and it will change the way you look at lottery games forever.

To give you a sample of the potency of LP System, let's look at one group. This is not among the most powerful group but will show you how it works. Your play should be accompanied with the money management chart. You can never know how much you win unless you know the cost of winning.

LP sample one

339

060

955

The key is to be patient. Like in the stock market you do not just jump in. You have to wait for the trend. You buy only when the stock is positioned on the indicators you are following. Then you ride the trend, stay with it as long as it is not violated.

When the LP lead number drop you take on the next two. Let's look at one place it dropped in New Jersey.

On March 20th 2009

339 dropped. That is the number you have been waiting for to start playing. You promptly take on the other two 060 and 955. The system skipped the middle game and dropped the 955 on May 3rd and again on May 5th 2009. Three weeks later it dropped the middle number 060 to complete that LP group.

If you had spent $4.00 every day on the two plays you would have gone through 36 days to hit both numbers. Your cost would have been 36 * $4 which would have been $144.00. Your winners would have been $1500.00 on those two numbers over the same period on the straight hit alone. The 955 dropped on 3/04/2009 in form of box 595 which would have netted another $160.00. As you can see the box would have paid for your total cost of playing ($144.00) and some left to play with. The $1520.00 would be all profit.

How many places can one invest $144.00 and get profit of $1520.00 on top of that. This profit came in only one and half month. If you can consistently hit the numbers like this you would not be looking at it like gamble. Most of the so called system out their even the authors do not know what they are talking about. They cannot defend it. The goal here is to make every subscriber a winner. The LP is from the greatest lottery master in the world. Coming from a family that owned lottery games outfit, studied everything you can think of about the lottery game both on English Australian fixed odds and the American lottery games. This is the only game that comes with monthly analysis. There is no system anywhere that will come close period. This system took nearly thirty years for completion. Try it and you would never go back.

When you use the LP system it takes you to where the winners are. You do not just play with your hard earned money as if you are throwing the darts and hoping that it will hit the big numbers.

There is no system in the world like it. Discard that old antiquated system that has been costing you money and play only when the number is set.

People compulsively chase numbers. A lot of people play that $1500 or more out of their pockets without even knowing it. The addiction do not give them time to seat back and ask how much they are spending and most often do not win. If you can wait patiently and play only when your number is set, then you will look at the whole game differently.

The LP system is the only book that will prepare you to play intelligently. If it takes me 36 days to play $144.00 and win $1520.00 after expenses you bet I will do it. Join the LP group and start winning.

Now let's use LP series 1 and look at one group in more than one state to show different ways the numbers can come. The goal always is to minimize your expenses and play with the house money whenever possible striving to hit the big winner. In worst case scenario look at not playing with huge sums of your money at the end of the whole thing.

We are going to run these through three completely different states to see how you would have paired using the unique LP system. We will run it through Iowa, New York and Tennessee.

The LP group is

619

886

733

The lead indicator is 619.

619 is the number you are waiting for. Once it drops you start playing the other numbers.

You will notice that the 619 dropped in New York on 4/21/2009. The middle number 886 played on 1/1/2009 and the last number 733 played on 6/08/2009. You will notice that in this case the game played in mixed order. The 866 dropped first followed by the lead number 619 and finally third number 733. Your target is to hit one or more of the numbers in the LP group and to do so with little or none of your money at the end. You go after the ones that haven't played.

It took about 155 days to play the other two numbers in exact order. If you had played average of $4.00 a day on those you would have spent about $620.00. On exact winning the straight would have been $1000.00 now you are in the profit margin. You have money in your pocket from the house after deducting your cost of playing not counting any that played in any order. In some cases the group may not play in any order (box) over that period of time. This happens in groups that are not among the strongest. The fact remains that if you come out ahead and not lose your money in the process you cannot ask for anything better than that. The LP system makes the lottery play almost take out the word gamble. Because the system meticulously set the groups, you should feel good when you play with the LP system.

We still have the other two states to examine on same LP group 619, 886 and 733.

In Iowa 619 dropped on 4/2/2009. The 886 played on 3/23/2009 and the 733 played on 5/7/2009. You will notice that the middle number dropped first on the LP group in Iowa followed by the lead indicator 619 in a little over one week and followed by 733 one month later. The two numbers on the LP group dropped within 32 days from the indicator. This would have been a bonanza for the member. In this case the member would have spent $128.00 played and the winnings would have been $1000.00 on the straight winners not counting any order (box) winnings. If you look at the winners in six way position you will notice that the number dropped sometimes. In situations where it drops within the time period you are playing the LP group that becomes additional profit to the above winners. Folks, it doesn't get better than this. In Iowa the LP number group played all three just like in New York except that it played in much shorter time thereby fetching more profit. This shows you that the group does not just play in one state which is why you watch for the lead indicator.

Now let's complete our examination of the same group in the third state Tennessee to see how it played. The middle number played on 3/25/2009 followed by the third number in less than two weeks. 733 dropped on 4/1/2009. Folks that is not an April fool. It came back two days later in exact order. The next number we are playing is the remaining one 619. At this point you have a choice of playing to catch the 619 or walk away with the $1000.00 you netted within one week. If you played the group in Tennessee you would have won $1000.00 after playing a total of $28.00 from your pocket. That sounds like a healthy profit from $28.00 investment. Only the LP system will get you this. If it is not LP, then you are taking a big gamble. Join the winners.

The key is and will always remain to be patient coupled with the system that will get you there. The critical question remains how much you are spending on lottery. So many players are almost always ok with losing and continue to play anyway. They do not have any system in place. They do not know why they are playing certain numbers other than the fact that they are going to play anyway. The fact remains that this is your hard earned money. You should treat it as such. We do not believe that you should play any number without trigger. You have to aim at something to have a chance. The other problem lottery players have is patience. When the odds of hitting pick three is 1 in 1000, it becomes very unlikely that you are not going to win the next game. What will be the likely frequency of winning? If you win based on the mathematical combinations of three numbers out of 0 -9, you have to spend $1000.00 to win $500.00. You can see right away that the scale is not balanced. It is designed to tilt towards the house by taking advantage of those things that you should control like patience. How do I develop the patience? How do I not play compulsively? How do I monitor how much I am spending? What is the likelihood that I could come out ahead? The answer lies in the LP system. It is like no other system out there. LP shatters the myth! LP destroys the odds! LP will get you there!

We consistently give you analysis of what happened when you use LP groups. We give you individual states and dates.

We in some instance give you more than one or even two states. This is clearly to prove to you that every number through LP has to produce certain results.

Now let's look at another LP group to see what happened. Remember the goal is not to play with large sums of your money as much as possible. If you can come out playing with the house money and eventually catching the big winner that will be worth it. Do not play with any system that makes you just waste your money with no end in sight?

If you see yourself sounding like "I missed it by one number. I was so close. Maybe I get them in the next game. If I select enough I might hit the winning number." If the system you are using gives you too many numbers and you just pick all or most of those numbers, you are wasting your money. In fact the system has no pattern and should not be called a system at all. Most of the so called competition is out there to get your money. Profit should not be at the forefront of anything worthwhile. Validation is the key! They do the best professional pitch that will make you spend your money buying their junk without even knowing it. We intend to make every subscriber a winner. There is no lottery guide ever used by any human being that will come close to LP. Each and every LP group you play has LP trigger. We do not advocate playing the number until the trigger shows up. The trigger is the green light. You have to have that to proceed. That is the beauty of the LP system.

LP number group will consistently show you that it has played in other states. Our unique system will isolate the numbers and group them exactly on how they are supposed to play. You pick the batch that is playing in your home state and stick with it. Play to win and not just to gamble.

Now let us look at a set of LP group to see what happened.

A	**B**	**C**
241	014	999
031	903	171
207	297	366

From group A

Virginia started from the middle and dropped 031 on 1/9/2009 before going to the lead number almost four months later to drop 241 on 4/8/2009 leaving the 207 for completion. We are talking straight hit here which will be equivalent of $500.00 for each $1.00 straight. Let us assume that this game took a long time to drop from the middle one. It would have been about 89 days from the indicator. On this long dry spell, if you had been spending 4.00 a day on the number, you would have spent $356.00 and the winner would have given you $580.00. As you can see you are still ahead and now playing with the house money to hit the other number to complete the LP group. After deducting your expenses so far you now have $224.00 pure profit. Anytime the number takes a long spell to come out the likelihood that the next one will drop soon is very high. If you play the 207 for the next 56 days and walk away you would have done this without spending $1.00 from your pocket. No other so called system out there can do this. You can now play with the house money to cover you for the next 56 days without spending $1.00 from your pocket. Folks that is how to play the lottery. You have to play intelligently. The reason is because of the LP system. We are going to test this group A in other states to see how it plays out.

We now plug the A group into North Carolina. The 207 dropped first on 4/10/2009. You now take the next two and guess what, the very next day the lead number hit 241 on 4/11/2009. It does not get better than that. You would have spent $4.00 and it would have netted you $580.00 the very same day. In this case you can chase the third number 031 with the house money or walk away with your $580.00. At $4.00 a day it will take you another 145 days to exhaust the house money before putting any of your money. You will be scooping the money from the box and straight winnings if the number shows up before 145 days. This is the group A that played in Virginia. You can see clearly that it took a different pattern in North Carolina. In each of those states you could have come out a winner without spending your money.

Let us assume that you are not yet completely convinced. We are going to run the same group in one more state to show you that there is nothing like the LP system.

Missouri:

The A group started playing in Missouri from the bottom up. It dropped 207 on 1/23/2009. Less than one month later the middle number 031 hit on 2/20/2009. In less than two months the 207 repeated itself on 4/9/2009 and what remains to play will be the lead number 241. You can once more see that you will be playing with the house money. Starting from 1/24/2009, you would have spent about $108.00 and won $580.00 to continue if you choose to. This obviously is happening in different states. I am sure you would like these numbers to be played in your home state. These winnings did not even mention all the winnings that come in box which would have netted you more money. In some cases the box covers all your expenses and the straight wins are all profit. If it is not the LP system, you could be wasting your money.

Let us take a look at the underlined LP group B. We are going to give you enough to cover different states at different times. The only advice is that you do not start playing a particular group until the trigger shows up.

Now, let's look at how the underlined group B came out.

014

903

297

We are just going to look through one state in this group. By now you know that the LP group appears in different states at different times. It has already been grouped with the unique LP System. Do not waste your money until you see the trigger.

Watching in the state of Tennessee the middle number 903 played as the trigger on 1/7/2009. You immediately take on the other two. Two weeks later the third number 297 dropped on 1/22/2009. Spending $4.00 every day on these numbers you would have spent $60.00 out of pocket and ended up winning $580.00 in just two weeks. You now have $580.00 to pursue the remaining number 014 for another major hit or walk away with your winnings. This bounty is not counting the box winnings that could come in between all this. If any of the numbers appear in your home state, you now have the other numbers to play with. It does not matter the order it came. Any one of those can be the trigger.

We are quickly going to wrap up the underlined group. We will take a quick look on the underlined C group in just one state. In some cases the LP group will hit one right after the other. In some cases they hit in a matter of three days or less. When that happens all you can do is laugh your way to the bank. Imagine spending less than $20.00 in three days and walking away with nearly $2,000.

We will revisit at least one of those, but meanwhile let us complete the underlined group C.

999

171

366

We will look at only one state. Again if the trigger appears in your home state, by all means pursue the rest of the group.

In New Jersey 366 dropped on 1/20/2009. At this point you take your shopping bag and follow the other two 171 and 999. In less than one month you get rewarded with 999 on 2/15/2009. You have spent about $130.00 out of pocket and got $1,000. The 999 would have paid out $500.00 on each $1.00 spent. You have a choice of walking away a winner or playing with the house money to see if you will catch the next straight hit. In another two weeks the 999 came back. Assuming you continued three months from that point 366 came back just like the 999. Imagine how much you would have won. This is not counting the box. You will notice that both numbers came twice. The remaining number to hit from the group will be 171. Based on the history, I cannot tell you that 171 will not come twice. This is a potential gold mine waiting around the corner. Note these numbers and follow anytime they come in your home state. Only LP can point this out based on our system.

We are going to touch one that came almost right behind each other. It could happen in your next play. After this example you will then have basket of groups to choose from, neatly worked out with the LP system. Anything less than LP system you could be taking a big gamble.

When the LP group hit close together it sure is a big bonus. It is important that you start playing the other two numbers once the trigger hit. There is no need to delay. If your group happens to be one of the ones that play right after each other you could be leaving money on the table. Each time that you do not win, the house keeps your money, so never leave the winner for the house. Your goal should always be to come out ahead no matter what.

Close LP group hit

A	B	C
714	085	420
891	845	875
296	551	606

If you look at the state of Maryland the A group started dropping right from the beginning of the month of July 2009.

All three numbers came during the day for three straight days. Can you imagine if you had skipped one day? 714 dropped on 7/1/2009 followed by 891 on 7/2/2009 and 296 on 7/3/2009, all on the same month. That is unheard of that all three numbers came within three days right after one another. The ultimate fact is that they all came straight within those three days. That is the power of close LP group. You would be walking away with about $1160.00 after spending less than $20.00 in two days. There is nothing out there that will come close. Any of those numbers could have been the trigger and it would not have made any difference. It does not matter whichever way the house is playing the game once they come inside the LP group, you pursue the other ones. Do not play just for playing sake. Do not throw away your money as if you did not work hard for it. In this case you followed the group just as we recommended and got rewarded handsomely.

We have tested the system through any angle you can imagine. If you follow the system as well as use the money management chart you are likely going to come out ahead. We intend to shatter the myth. We intend to defy and blow away the conventional odds. We intend to make every subscriber a winner. We have consistently shown you what the LP system can do. Make sure that you collect all the series and not leave any money on the table.

Grab it while it lasts!

We have people out there that are playing with their retirement money. A lot of them will send their grand children sometimes across state lines to play for them. A lot of them are patient except the fact that they could be playing the wrong number. If you are patient and can play same numbers, the right numbers consistently, LP system is for you. Why play with your retirement money if you can play with the house money?

Watch out for your trigger number in the LP series before playing your money.

LP Series 1 Basket

031	077	094
207	142	767
903	033	258
058	099	057
547	017	370
957	436	686
081	036	028
401	355	818
015	861	013
024	025	062
896	247	850
833	969	512

Remember that the first numbers are the trigger. When you are watching out for the trigger, do not forget the other two numbers. If the house chooses to make the middle or the last number the trigger, you immediately choose the other two. It is as simple as that. The LP groups have meticulously been worked out. The above basket group has the lead numbers starting with 0(zero). It is intended to simplify the games for you. If the game you are watching begins with 0 you do not have to hunt everywhere for it. It has to come in exact order. We do not really believe that you are a winner until you start hitting straight with the LP System. The system is going to encompass triggers from 0 to 9. This is the LP series 1. Collect all series for the serious player that is aiming to win serious money. Grab it while it lasts!

LP Series 1 Basket cont.

171	146	183
366	109	687
338	776	852
143	197	128
393	865	400
157	336	319
145	103	103
735	128	035
869	757	286
156	186	127
368	133	176
550	566	852

You can see in rare cases when the trigger number is the same like 103 above. The likelihood that one of them will start in reversible order or from the middle is high giving you the direction to follow. If you decide to play both groups your chances and expenses will be in same proportion. Remember we strive to hit the winners straight. All the series are designed to cover enough grounds for those who want to play more than one group and remember that patience is the key. We have now given you lead triggers starting with 0 and 1.

When you get into pick 4, quinto, pick 5 and lotto then we are talking about probability. You will need a minimum of $10,000 to start dreaming of winning. When you talk about pick three, there is only pick 3 combinations of 1,000 numbers of 0 through 9. We are not worried about the odds by any means. We have shown you countless times our LP System hit the numbers straight and put you way ahead of the competition. The fact of the matter is that there is really no competition. Grab the series and do not let go of any of them no matter how nice your neighbor or relative is. We intend to make every subscriber a winner.

We are now going to continue with the LP series 1 basket. The lead trigger now will be the number 2. This is to save you the trouble of scouting through every part of the book. The lead number will show you where to search.

LP Series 1 Basket cont.

241	297	258
031	999	362
207	171	569
205	254	253
554	932	764
695	842	548
262	201	203
280	143	920
520	393	701
229	286	228
099	002	869
017	192	028

There is this group of friends that drive taxi in Washington DC. Their big bond is playing the lottery. They play every day and continually look forward to playing the next game. Each time the result came out, you hear the regular sigh of "oh! I missed it by one number." Their job is one of those jobs you happen to make cash every day. I know one of them; Jay Indira has a wife with a very good job. I really wonder if he ever gives the wife money for feeding. He has a very big bag filled with old lottery tickets which he hopes to write off when he wins. He has done this for at least the past 15 years wagering minimum of $40.00 every day. He never seems to notice because of the fact that he makes cash. His biggest problem as well as others in his group is that they chase the numbers. When you chase the numbers often times, your play appears very close to the winner giving you renewed energy to keep on throwing your money away. Can you imagine playing 10 LP groups consistently on that $40.00 daily? Do you think that he would have been on a 15 year dry spell? It is your hard earned money and you should treat it as such. He practically lived to play the lottery every day. Anytime you see him, his head is always bent down copying numbers he will play in the next game only to come out disappointed once more. Because they make cash everyday no one seems to take note of how much they play on daily or monthly basis. The LP System will give you the numbers to play as well as the money management chart to know exactly how much you are spending.

LP Series 1 Basket cont.

366	310	385
338	077	363
736	142	414
336	362	355
756	569	469
963	638	094
384	326	392
356	428	470
764	652	859
393	336	319
157	589	679
992	643	619

Every group is neatly laid out for you. We do not believe that you have to spend all your time chasing the numbers or searching for it in our series. We have now gone through 0 through 3 of the series 1. When you collect all the series there will be more ample groups for you to play at any given moment. The goal remains to make every subscriber a winner. All you need to do is be patient and follow the simple instructions. We are now going to continue with the rest of the group.

LP Series 1 Basket cont.

425	469	497
336	094	725
756	767	876
428	414	481
652	310	583
481	077	471
470	436	464
859	933	852
622	573	128
400	438	471
319	228	579
679	869	975

We have just completed the LP group leading with number 4 and the remaining will be 5 through 9. When you look at the last of the group starting with 471, you will notice that the other two of that group is 579 and 975 respectively. When you look through your home states past results, you will notice that those numbers tend to repeat. They do repeat as well as the 471. In some cases they come more than two times within the same time frame. The LP System is so unique that it actually isolates those and have them in a platter for you. There is no other so called system out there that can do this.

When one group is different it is still in the LP group. The benefit is yours because the LP actually isolates them in that order.

If you go back few pages you will remember the rare instance that the group came right after one another In Maryland. That came in the month of July 1 to 3, 2009. Those numbers are 714, 891 and 296 respectively. That is from the A group. The next LP group beside it is the B group. The numbers on the B group are 085, 845 and 551 respectively. Once more if you go to your home state past results you will notice that this group tend to repeat. The unique LP system is the only tangible system able to point this out. When those groups appear in your home state, that is any of them as the lead trigger, take your shopping bag and start playing. It's like buy one get one free. There is no other way or a better way period.

We are going to continue with the rest of the group knowing that by now you are eager to have it all. Do not leave home without your LP book because you could be leaving a world of good.

LP Series 1 Basket cont.

569	520	587
638	355	625
835	469	693
554	565	547
695	648	957
949	058	497
579	548	583
975	326	471
253	428	579
539	520	520
647	814	392
485	839	470

Again you can see that the last two of the group started with the 520. Do not forget what we told you before in cases like this. This is two competing groups with similar lead trigger. One of them could start in reverse order or the middle. The starting point determines which one you want to play or if you want to play both. Playing both cost you twice as much and your benefit is double. Nothing here negates the fact that the LP System is designed to give you major hit on any of the groups. We do not believe that you are a winner until you hit the winner straight.

LP Series 1 Basket cont.

668	687	638
385	852	835
363	635	876
635	695	693
146	949	384
109	672	356
648	652	647
058	481	485
547	583	262
619	658	657
886	464	365
733	852	896

We have now completed the LP system through the lead trigger 6. Remember that it can start from any of the group and you just follow suit. Do not play like Jay Indira. The herds get slaughtered. Insanity is doing the same thing and expecting a different result. Do not continue to use the old antiquated system. Check how much you have spent using that so called system that has been costing you money. The sales pitch is professionally done. The book cover is enticing. The language is psychologically done to get you, the reader to buy. The end justifies the means. If you have not been winning straight, do not stay with the same book. If you win box once in a while, it is not enough to justify buying the book that does not work. Some people win box and feel very good. It renews their hope of winning straight in the future. They never really check how much they have spent. In fact they play twice their normal wager around that period. They spend the winning in a matter of days and back to square one. The fundamental question remains, how much have you spent since the last time before the little win?

The right system would be more than glad to show you how much wager is costing you or how much you are spending in a given month. The reason is simple. We are not spending our time giving you silly sales pitch. We spend our time giving you enough to make sure that your winning ought to be big. The real winner ought to be straight period. Anything less than that you are taking a big gamble with your hard earned money.

Now we are getting to the completion of the LP series 1. This will give you a wide array of choices within the series 1. To have a lot more choices, collect the other series 1 through 9 to have enough at any given moment.

LP Series 1 Basket cont.

797	742	757
957	922	756
274	285	743
745	778	769
634	595	962
302	594	784
785	795	767
951	881	521
941	283	603
714	756	757
675	134	156
213	343	262

In the first part of the 7 LP group starting with 757 as the lead trigger, the next number shows difference 1 showing 756. Now you have 756 following 757. Naturally most people that chase numbers or just play anything and everything in front of them will play 755 thinking that it will continue playing difference 1. You will notice that the third number to complete that group came as 743. The LP System does not miss a beat. It clearly shows the subscriber that the next game is 743 and not necessarily difference 1. The house will not make it that simple. Our unique system follows the number wherever it goes.

Coming down on same LP group with 7 you will notice that the first group of the third row showed similar pattern. The lead trigger is 785 followed by 951 and 941. You will notice that the middle number is 951 and the third number is 941. Naturally the first number is supposed to be 961 but the LP system is able to point it out. The lead number is 785 which is quite different from 961. If that group is playing in your home state and the first trigger played as 785 followed by the middle number 951, the LP automatically shows you that the group is positioned to play difference 1 which will be

941 despite the fact that the lead trigger 785 appeared like a decoy. There is no other system out there that can do this. To be a winner you have to go with the best. In life if you stay with losers that is what you are going to get. The LP system is the best, hands down!

The LP Series 1 cont.

844	850	861
561	410	745
481	103	634
818	896	877
013	833	059
156	337	759
888	850	885
228	512	676
606	417	795
881	852	810
283	760	834

062   958   629

If you play your home state lottery properly you could come out a very happy camper. Do not play just for playing sake. Do not believe that you are going to lose like before. You have something different. You have the system that is the best all over the world in your hand. You are better equipped than the regular neighborhood lottery player. Some folks play, drink liquor, lose as usual and repeat the same thing. You should be conscious of your money. There is absolutely no reason for you to play if cannot come out ahead. The time you pick the numbers, drive to your local store, burn your gas, queue on the line to play, should be enough to justify winning. The person with the right formula wins consistently and the person with the wrong formula loses all the time.

LP Series 1 conclusion

957	975	922
274	574	285
692	742	441
902	969	962
024	332	784
896	376	956
948	951	958
255	941	810
479	488	834
965	999	926
850	879	692
714	334	415

LP Series 1 Triple Stars

You cannot complete this book without the triple stars. Every player wants to get those triple. They are the easiest if you catch them when they are playing in your home state. We have them in the lead trigger position here. If the game starts playing from the middle or the third number you will notice it as a good observer. We enclose herein the triple stars of the series 1 group.

LP Series 1 Triple Stars

000	111	222	333
815	204	332	564
596	969	709	009
444	555	666	777
032	637	221	776
533	516	384	713
888	999		
216	923		
085	022		

This is the completion of the LP system series 1. Grab the other 9 series for the serious player that wants to win serious money. You have the best system ever created in your hand. Do not play like Jay Indira. Besides wagering a lot of money blindly he could be causing himself health problems. He sits in his taxi and gets immersed into picking the lottery game blindly. Over the years I knew of only one time he won $500.00. For somebody that plays an average of $40.00 per day, that is not a winner. He spends more than that in one month playing lottery. He sits to a point that he does not even care about picking passengers when he is too deep into the game.

How do I know that this is the best system in the whole world? In lottery game, I occupy a place where most people can only dream about.

Overseas lottery outfits are owned by individuals. My father happens to own one of those. He had publications in this business while I was growing up. I used to go to his office to work during my high school holidays. A lot of people came to the office to play lottery. They were mainly playing the English fixed odds or what they call the English Pools game. Most of the people that play daily are the laborers. They work so hard and end up spending a lot of their money hoping to win the next game. I got worried watching them play and lose all the time. I started questioning the logic behind the numbers. I disturbed my father to the point that he showed me how the system works. I embarked on breaking every logic known to man behind each and every lottery game. I mastered the English system and continued my quest on the American lottery games. I have never been in a hurry to share what I know until I got to the point I have tested and answered the question.

The key remains that every lottery number must travel in certain direction until altered. When it is altered how many people can honestly recalculate the new direction and follow it? This is what separates me from the pack. It took fine tuning of nearly thirty years to say yes I can offer this to the public. My satisfaction is that you win. I am not talking about winning box. With the LP System you are never really a winner until you win straight. Each and every group is painfully worked through the LP rigor with you the subscriber in mind. There is nobody anywhere that can come close when it comes to lottery guide. The lottery number is beyond just looking at the numbers in front of you.

When you have the book from the best, you are just like the best. We strive to make you a big winner period. Follow the simple instruction, be patient and believe me, you will come out a winner!

The LP System General Market

For those subscribers that look to have wide array of choice. Each group has gone the rigor of LP System.

870	433	661	420
175	198	752	185
566	807	544	645
114	490	770	885
717	978	127	474
785	785	481	654
091	837	458	030
893	125	545	807
854	149	570	702

596	581	650	615
291	687	880	887
258	459	567	605
654	740	587	643
676	199	017	019
567	155	656	580
666	542	711	949
875	618	071	897
054	164	078	478
077	388	109	083
512	847	289	512
648	765	485	514

445	003	059	558
754	880	729	468
557	570	125	445
165	761	865	074
388	588	567	719
456	960	056	215
658	164	066	554
101	701	087	761
865	958	705	516
471	094	707	338
607	989	751	984
707	547	664	076

610	205	144	900
528	851	175	788
448	451	855	857
705	277	649	035
272	519	602	354
812	711	717	384
729	249	697	477
179	253	643	259
537	767	729	374

Follow the LP System instruction and I guarantee you will always come out a winner.

The LP System is by the best lottery forecaster in the world.

Winning Every Pick 4 Game.

So many people that wager on numbers have had situations when they played pick 4 and missed by one number. The common saying is that the house played the number by subtracting or adding one. They actually feel like the house saw what they are playing and chose not to pay anybody.

If your state result was 2165 followed by 8480 while the neighboring state played 2165 and 0333 the rest of the games in the two group will be completely different.

Take a look at the rest of each group to see that they created different paths.

Group 1A

2165

8480

8003

0133

3302

7113

8406

0591

Group 1B

2165

0333

9081

2092

5753

9677

7061

0727

You can see that each group produced different sets of pick 4 numbers. The numbers that looked close are the 0133 and 3302 from Group 1A and 0333 from Group 1B.

If you are playing following Group 1A and they played 2165, 8480, 8003 and the fourth game played 0333, you have to shift your game to Group 1B. The bettor without Lottery Pal will believe that the house played 0333 when it was meant to be 0133.

The above two tables clearly shows you that they are playing two different trends despite the fact that both started with 2165. If the game played 0333, you immediately need to start playing 9081 among others. This is the reason why pick 4 is probability.

If you have Lottery Pal series, all you need to do is find the number and follow the trend.

The numbers in any group play at different dates because each successive game might come from a different group as in the above case. The introduction of that game changes the entire equation. You are not going to find many people that understand how to do the calculation.

The Lottery Pal is here to address that problem.

So many people take the potential payout for granted. The pick 4 game pays upwards of five thousand percent on any straight win of one dollar. You cannot expect all the numbers in any group to play right after each other. it is not going to happen. The closest you can come to that is by identifying the winning numbers through Lottery Pal series.

You can, however, identify any group like the example above and follow the trend. If you think that pick 4 games cannot be won, think again. All you need is the collection of Lottery Pal.

Let us take a look Group 2

8555

8556

6521

6526

0161

2001

0281

8005

6208

6860

You will notice that DC 4 lottery played 8555 and 8556 on 1/1/2009 to start that year. The 8556 that played on one month anniversary is closely related to the 8555. Take a look at the rest of the Group2 and the dates they played.

DC 4	Dates Played
8555	1/1/2009
8556	2/1/2009
2801	1/2/2009,
5261	4/9/2009,
6608	8/30/2009,
8206	11/10/2009
2656	5/29/2009,
5800	8/20/2009

1002   8/29/2009

0116   12/19/2009

As you can see from the dates, the numbers in this group played at different dates over the whole of that year. The entire group is not yet exhausted at the end of that year.

This means that you have to meticulously follow the trend. In this case for a period of more than one year.

The second option is to identify the next number in the Lottery Pal series and follow that trend. The series saves you the trouble of waiting. The calculations are already made.

You can run Group 1 that started with 2165 through different states.

I will give you three other groups for your records.

Group 3

0426

0455

5947

7563

2505

6946

9563

1201

Group 4

6279

6579

0320

3471

0933

9903

4593

6592

Group 5

4921

5573

9958

6189

8553

0476

7672

4298

The above groups are represent each trend. They are not going to play right after each other. Lottery Pal series 2 will begin to go deeper into the world of pick 4.

You can win those five thousand dollars consistently. The only think you need to do is follow the Lottery Pal series. The winning numbers are set in simple order to make it easier for the readers.

Go through the rest of the book and start winning every day.

There is no lottery book that can come close.

The next Lottery Pal series will begin to do exhaustive breakdown of every pick 4 and pick 3 trend that you will not find in any other book.

All rights reserved. No part of this material shall be reproduced or transmitted in any form by any means, including photocopying, recording without written permission from the publisher. The numbers in this book has been carefully worked out with the rigors of Lottery Pal and our readers in mind; however, these are recommendations only.

Made in United States
North Haven, CT
31 March 2024

50731153R00043